The American Poetry Series

VOLUME 1   Sandra McPherson, *Radiation*
VOLUME 2   David Young, *Boxcars*
VOLUME 3   Al Lee, *Time*
VOLUME 4   James Reiss, *The Breathers*
VOLUME 5   Louise Glück, *The House on Marshland*
VOLUME 6   David McElroy, *Making It Simple*
VOLUME 7   Dennis Schmitz, *Goodwill, Inc.*
VOLUME 8   John Ashbery, *The Double Dream of Spring*
VOLUME 9   Lawrence Raab, *The Collector of Cold Weather*
VOLUME 10   Stanley Plumly, *Out-of-the-Body Travel*
VOLUME 11   Louise Bogan, *The Blue Estuaries*
VOLUME 12   John Ashbery, *Rivers and Mountains*
VOLUME 13   Laura Jensen, *Bad Boats*

BAD BOATS

BAD BOATS
Laura Jensen

The Ecco Press
New York

Copyright © by Laura Jensen, 1968, 1969, 1971, 1972, 1973, 1974, 1975, 1976, 1977
All rights reserved
First published by The Ecco Press in 1977
1 West 30th Street, New York, New York 10001

Published simultaneously in Canada by Penguin Books Canada Limited
Printed in the United States of America
Designed by Harry Ford
The Ecco Press logo by Ahmed Yacoubi

Library of Congress Cataloging in Publication Data
Jensen, Laura, 1948–
    Bad Boats
    (The American poetry series; v. 13)
    Poems.
    I. Title.
PS3560.E59B3   811'.5'4   77-71289
ISBN 0-912-94639-3 clothbound
ISBN 0-912-94640-7 paperbound

Grateful acknowledgment is made to the following magazines in which these poems first appeared: ANTAEUS: "I Buy Nothing," "Sleep in the Heat," "Indian," "Tomorrow in a Story," "Water Widow," "She Will Not Dress Herself," "Faces Passing Your Garden," "The Poorsoul," "Vespers," "Probably the Farmer," "The Complex Mechanism of the Up." CONSUMPTION: "The Red Dog." EUREKA REVIEW: "Redwing Blackbird," "Bad Boats." FIELD: "After I Have Voted," "Landscape," "An Age," "Praise," "The Sparrows of Iowa," "Dreaming of Horses," "The Cloud Parade," "Night Typewriter Sounds," "Patience Is a Leveling Thing," "House Is an Enigma," "Winter Evening Poem," "As the Window Darkens." GROVE: "Actors in the Country." MADRONA: "Talking to the Mule." THE NEW YORKER: "Animal," "In the Hospital," "Writing Your Love Down," "Heavy Snowfall in a Year Gone Past," "Baskets." OPEN PLACES: "All of Them Who Ran from Rainclouds," "The Harvest," "The Crow Is Mischief," "Amigo Acres," "Statue Maker." POETRY NORTHWEST: "The Ajax Samples," "The Only Dream." SENECA REVIEW: "Happiness." Some of these poems also appeared in ANXIETY AND ASHES (Penumbra Press) and AFTER I HAVE VOTED (Gemini Press).

Contents

1

The Red Dog	3
Talking to the Mule	4
Conversations in the Primate House	5
The Ajax Samples	6
I Buy Nothing	7
The Harvest	8
After I Have Voted	9

2

Sleep in the Heat	13
Animal	14
Indian	16
In the Hospital	18
Landscape	21
The Only Dream	22
Subject Matter	23

3

The Poorsoul	27
Actors in the Country	28
An Age	29

Vespers	30
Praise	31
The Sparrows of Iowa	32
Dreaming of Horses	33
Redwing Blackbird	34
Probably the Farmer	35
The Complex Mechanism of the Up	36
The Cloud Parade	37

4

Heavy Snowfall in a Year Gone Past	41
As the Window Darkens	43
House Is an Enigma	45
Winter Evening Poem	47
Happiness	48
Amigo Acres	49
Water Widow	50
All of Them Who Ran from Rainclouds	51
Statue Maker	52
Faces Passing Your Garden	53
Patience Is a Leveling Thing	54
The Crow Is Mischief	55
Writing Your Love Down	56
Baskets	57
Bad Boats	58
She Will Not Dress Herself	59
Here in the Night	60
Night Typewriter Sounds	61
Tomorrow in a Story	62

(1)

The Red Dog

You know that he is going to die
as soon as I tell you
he is standing beside me
his hair in spikes and dripping
from his body. He turns his head.
Canadian geese
all of them floating along the shore.
The red dog is swimming for them
only his head shows now
they flap into a curve and move
farther along the bay.
You know that he is going to die
this is the time for it
this is the best time for it
while there is a way to vanish
while the geese are moving off
to be their hard sounds
as their bodies leave the water.

Talking to the Mule

In the evening there is a snail
passing through life across your lawn,
his trail official, a government seal.
The lawn is delicious, dry and cool,
the color of pepper.

Someone took a gun and shot that cone
off the pine tree.
He might have stopped to think
you heard for weeks the way the wind
blew through its sculptured sections.
It had skin soft as your ears.

The stars have started out the night.
They are something of importance
that clouds cover. They are too much
to be seen, but still they belong to us.

Rub your nose along the fence.
Tip back your head and bray,
for night is yours. It is never
against you. You are not its enemy.

Conversations in the Primate House

I am betraying a white horse
standing beside a Polish store.
But that horse was my betrayer,
like the moths it beat on your window screens
in autumn, deciding your future.

You never betrayed me, sister.
That was the year we were possessed.
Then there was wind and dogs,
and we were running
and the streets were falling
in the gardens.

I am betraying you, sister.
I see ways you tried protecting me,
extending me time in the godlike stature.
But we are born together every year,
with wide dark eyes and narrow hands.

I am betraying my keeper.
He could not know that I half follow birds
as they fly out among the trees.
But that is the story of our heavier bones
and the other charted mysteries
of the closely grazing ones.

My sisters fed on leaves when they were young.
We are fed on fruit and paper
and stretch our arms out on the swings.
I betray my sister when I see her,
when she opens her wings.

The Ajax Samples

They gave us the mysterious deep warehouse
filled with lavender and telephone books.
We emptied it.
I gave the last box to a beach house
on Brown's Point.
It was like running down into a stadium
crowds cheered in the blackberries on both sides
there was a wide blue field
with a sailboat anchored
the house with its shingles and screen door.

They gave us Alan Gentili.
He was our inspector.
Alan Gentili how beautiful your eyes are
your thin brown legs in your khaki shorts.

They released the pigeons over Melrose Street.
The pigeons turned together in the afternoon
again and again
and their shadow was a blessing.

It was like a sign to us.
It was like starting our lives in another way.
Saint Nicholas, the tooth fairy
they have nothing on us now.
We are so glorious that halos shine around us.
Jewels glint on Alan Gentili's panel truck.

We are the Magi traveling by moonlight.
We are Michael Anthony
giving million-dollar checks to strangers.
We are the holy sunlit wanderers
and we leave gifts behind.

I Buy Nothing

1
I buy nothing. I am nothing
like a sailboat just touching water
carrying a wake behind, I turn
like air on air.

We crushed wisteria and camellias
to make perfume, thoughtless
as a scar or as the sea,
petals larger than our hands.

Across the summer grass I walk,
a substance in the sun
that leaves a long black shadow,
truly, as if I lived somewhere.

2
The stars shine dimly on the town
limiting colors in the grass and leaves.
The fences fall between us;
I am not like you, or leaves or stars.

Salt falls from the sky into my clothes.
There are soldiers in the roses.
Everything happens after dark;
night blooms; the owls do not turn away.

The owls listen after dark
without blame or wisdom, resting
just above the roses. The tree
receives them, holds them in one hand.

The Harvest

There are no mistakes.
Before, he really was the czar
walking through the arrowheads and medals.
He was having a bad time, you say.
You say there was no music.
You say his coins were tears, and the curio shop
was more than he could bear.

The waiter was afraid to fall
into the hands of other people.
He brought the bottle for the czar
and looked at him with sympathy.
When it was night
he turned up all the chairs.

These are old photographs.
There is the czar,
there is the soul of the czar,
there is the waiter,
standing in the forest without faith
or motion, every moment
a carefully made decision.

What was it like to stand beside him?
I am a soul without a body.
I put on my shoes
and walk through the trees.

After I Have Voted

I move the curtain back,
and something has gone wrong.
I am in a smoky place,

an Algerian café.
They turn the spotlight toward me;
the band begins to play.

The audience stares back at me.
They polish off their glasses.
They ask the waiter, "Who is she?"

He holds his pen
against his heart.
He speaks behind his hand.

There are tea bags swinging
from their mouths.
Their teeth are made of brass.

The Jell-O sighs into the candlelight.
My eyes turn into stars.
Ah—the colored spangles on my clothes,

the violet flashlights and guitars!

(2)

Sleep in the Heat

I switch on the light. Crickets tick,
and the clock hands grow together, no record
of their own nocturnal repetitions.
Some things, for instance, branches,
can recall their circle
in the wind's big silent whistle,
their circle and returning touch.

The dark is dizzy. Within the shade
of heat, this stubborn demand for sleep
is slender. I try to please, I think
of hearts, their shape like lilac leaves;
I try to balance—one sheep fills me,
one is a shapeless chance,
one disobedience, one regard.
They feel I do not deserve them;
they are sleepy and kept up all night.

The sheep have hunger. Slowly they fade
into my eyes. My breath is their noon whistle.
Waking they are in me,
grazing in the pastures of my tongue.
It is morning and I brush them out.

When the tricks have all worn out
it will be winter.
The rain will replace the rage
of the sizzling crickets.
I know this—I am looking back.
Heat does not deceive me; when the rains come,
they will not blame me for anything.

Animal

I live beneath a house,
a large animal
no one can name.

I know the owl wakes in the sunset
and his lenses do not change.

I have heard the cat
that steps with itself
from room to room.

The bear in his own den
protects what he has claimed.

All animals are domestic.
Even the snake
bakes on his own rock.

Nothing is out of place.
Nothing inappropriate.
Nothing is too thin
to endure the danger.

I am a cousin
of the crow's collection.

I am keeping
for the girl her tea
for the dragon his silver
for the mouse his grain
grass for the herds that graze.

I am that other animal,
comfort. Everything,
even the winds that threaten,
come to me.

Indian

1
In lungs fresh like honeycomb
my fiery bright breath has trundled
victorious, a forest full of fire,
plunging as one with the bear.

And I am the cover girl,
the vestibule neck above the satin,
chiffon hands that dine and dance
under scented awnings.

For my heart is like an orange,
my teeth are white again like painted gates,
my neck of hair entirely feathers,
black hair adorned with petals.

I am swan flower, winged rose,
I bring my language rightly to my tongue
(where sweet genetics veer like dragonflies),
its battles distant and clear.

2
I watch whim, with narrow eyes,
fly true into darkness like an arrow;
buried in the sand, lit by a fire, past
becomes crime, branded by haste.

Dogs growl in their dark leather,
and ducks ask, loosening over water;
with the Navajo blood and turquoise rings
I look out soon as moonset,

a fluorescent skeleton,
steps down the cliff of this lighted building,
the burning and brick-red and desert rose,
to the dark, certain people

under a room of banners
which are a coarse black flag few people name.
I am entrusted to the ignorant prejudice—
loose my braid to fly my hair.

In the Hospital

Night breaths, short ones
from the garden, soft gasps
of snapdragon. Winter
reminds me of flowers.
Breath, where are you traveling?
You are not a caretaker,
but an illegible flower.
Winter reminds me of flowers.
Breath, the follower of air.

Night. Snow.

You forgot geriatrics,
anesthetic pillows,
taking back the seagull,
the spiritual moments of air.
Breath was what you lost
of her, that last moment.

Her love returns to you,
moment by moment, in the oval
messages of snow, in the oval
lips of cups. The curtains
are a message of her veil,
the glass the thin wall
of her clothes, the drawers
like the lives who used her cradle.

I look out my window
at the bent man and his
frosty plumes of breath
which explain a wealth of sight
within his dark coat, within
his cold shoes and questionable
books. I look at your tears,
the white cinders of snow
that the earth is carrying.

At this moment the reason
for your breath is given, the reason being
truce between bones
and the skin of your hands,
the truce of these dark trees
on the clean sheets of snow.

A thought at a time, a world
at a time, like the sky
without another meaning, no matter
how you look, no matter how
you lower your eyes
from its opening heart of color,
an answer to an answer to a prayer.

Mother, I speak softly,
and Father does not hear.
The times he has cupped his hand
and shaken his paper
are a welcome from your mother,
who now sends me running and grasping
through the invisible cells, into mysteries
of shapes and existence, our lives

tumbled into us like a cord of firewood
burning, the pain forgetfulness,
the message a puzzle like one of the cups,
Made in Occupied Japan.

I have been embarrassed by the testimony
of the tongues, changed roads, habits,
of the moment of prayer
for the opening of my heart,
the pain and swelling.
Didn't I always understand
the snow and the laughter, the bay leaves
of the cupboards resting in the snow?

Landscape

Nothing can oppose the cloud.
Nothing can oppose the gray
that sponges up the rust
off the old grass,
unless it is the stone
of its own color
in the tower
where the windows webbed over
are less open than its padlocked door.

It is not the gray birds
it is not the talons of birds
it is not the weather
or the trees that play dead
or the gray eyes of an old woman
or the children who are watching the ground
for sticks.

See what is coming—
a landscape where we take in turn
what is bleak and empty.
You do not comprehend yourself
until someone steps to you,
grateful you are carrying that lantern.

The Only Dream

It was morning and she found herself
in the orchard. The sunlight
shifted rich as cider in the winter
and she flew from tree to tree
in her long clothes, pulling jay feathers
from her tangled hair. There were so many
of them, it took so long to pull them
from her hair.
She rested there behind an ancient tree.

You are a witch. She laughed, her face
contorted by the tombstone. *No. Not that.*
You are a witch. She frowned. *Not that.*
She was a witch.
You are yourself at peace, like that.

This was the orchard
with her head above the grass
looking around at the umber trees
white in places in the lichen sunlight.

Subject Matter

On the good day, it cracks out,
the recognizable fowl that falls in love with you,
nothing to offer but itself in your eyes.
When you keep walking it starts after, helpless,
unrejectable. It would never harm you, you are certain.

Someone has seen one go fighting, taking a chevron
in lieu of the humming garden. Roses are red
because their ears are burning.
Someone hears the ocean and repeats its sound.

In your favorite country, the one no one remembers yet,
the hen, the duck, the other ones that settle—
all of them are minor deities. It is with composure
that they see flocks turn to the other mothers,
the peculiar foreigners they look in the eye.

(3)

The Poorsoul

The poorsoul steps from its habitat of leaves,
and it must be singing a song. Without the song
you would walk away, chanting Onlies
to Only the Wind, Only Little Sighs by the Water,
Only Your Neighbor.
Soon you will be leading every traveler
to the right saloon.
The poorsoul steps from its habitat of leaves,
and there must be snow. Without the snow
you would never say, "There, I thought it was
something living, and it has no shoes."
There must be summer, when the nights in gloves
wrap days around the feet of the poorsoul.
There must be a song, and snow in summer.
We would never stand there long in the cold.

Actors in the Country

Visual aid: a greeting card that expands
to an accordion of flowers.
The card is cut and planned to spread
to a flourish of green nasturtiums.
As the card is open, when it is open,
so each branch is open like a fan.

Under the trees,
what nonsense, someone traveling
all this way and into a theater
to meet the one traveling halfway,
in an ear! But try to find it right,
for the cloud falls like a monument
across the farm community.

It falls open, broken from the core;
stones break from the heart; are broken.
The rain, the dust, are only the same
as parts of disaster; they nourish history.
Meeting and parting like the ocean
as volunteers for counting,
oh, they do vary: swans and gondoliers.
Where they appear the next time they sleep,
as the expectations are opened
in their opening
words, country people, whom else
would they ask to break within?

An Age

There is a growth that hurts the child
one was, the child who still knew
the ocean rock from a distance,
flocked like cloth, white like sugar,
a flower out of focus in the waves,

the waves, the thousands of horizons
seen again and again as blue Japan—
something that changes the freighters
twined with lights and evergreen
in the port of Seattle.

I remember being nowhere in the early light,
halfway over ocean to a northbound freighter
and walking back to my sister
where our white wood caught fire
in the white sand.

Vespers

November

You are meeting me as I have come home.
I come home this way, through the watercolor
of the evening. Maybe you know the place—
where the south and the west
have the northeast as guest
and I am the cook who serves them. We all
have chairs.
I am telling you
of weeping memories from the sky; that tears
destroy you or shake you into being.
As you can see
I have brought friends.
There are books, paper, the Christmas cards
with the right design, some narrow poems
breaking down to sand.
We all make mistakes. They knock me unsteady
like a hand stepping down upon the service bell,
like a clerk turning.
One no more takes flight
than one must take one's flight
as an example. No wings, no;
but for the blossoms
that could sing now down into the stones
a sense of disaster. Rain, rain,
there you are.

Praise

What is praise but a terrible storm
you manage through?
I think they begin to appeal to you,
like talented mornings, like sparse good-byes,
the things that are best
for you: the wide hat you have affected,
your crop of snails, the broken leaves
you always thought could be repaired,
and the glass rain on top of your shadow.
Next door it is filling a claw-foot tub,
and falling on a bottle.
The sleep like a lowering shade,
your private book, the mighty crossing
of the aircraft over your pores,
an ambulance, the brain of the town,
they try to say, "I am so proud of you . . ."

I want to respect the sail again,
to believe that the wind moves slowly,
as slowly the sail fills. I want to know
that the sail has not come to an end
in uselessness; its joy should not end.
I want it to be sometimes as I have said,
not running and joking, and when it is up
and around like a flag on a staff,
for only a moment it imitates . . .

The Sparrows of Iowa

There is a secret in the miniature grass
that the sun will fall again and again
and the bats will weigh onto the neck,
tremble to the hair,
and if it is not you it will not matter.
It will be me, or one of the cattle.

It will be an answer for the sparrows
of Iowa, listed there as if no more exist.
They have been long with all of us,
chattering the bushes, ponderous,
and never been vermin. Their legs
are the dry bit you snip absently
from a houseplant—much better to say,

    once there were roses on the horns
    and thorns of reptiles, and they were
    the birds to originate warmth near the heart—
    they were the notes on the flute—
    the innocent liking for crumbs
    the day
    the percher on the chimney
    the smoke that goes out to look.

Dreaming of Horses

> *Etymologically, the word "mare" of your "nightmare" denotes a beast that is not a female horse.*

The hay is sweet for the partisan horses.
They helped you from the rope of your mother,
came shy to the lover dogs and your wife,
mammal to mammal, ran with straight backs
through the round forest.
Dreaming of horses the harts ran,
stark as the trees. Dreaming of horses
you open your door, and they are horses
standing in light, not beasts of the sound.
Your nightmares are no feminine clovers,
says an open etymological rose.

In the stables, a supporting opinion:
"It is like waking one day years ago
as Sister Radiance opened the shutters
for the last time; and for the last time
she brushed my hair and listened to me pray.
Mother the Queen gave me seeds for the birds;
Sister Radiance packed her belongings
and I walked with her, that first day.
I went, all grown, to matins after that.
She hugged me and we parted under bells."

Redwing Blackbird

Yes, a locust drilling, yes, its dwelling;
how did I wish to be a little sad?
Somewhere a tree is weaving a basket.
Somewhere a cow is drinking rainwater.
Where are the tracks?
                       Somewhere a train leads
health through the promised land of perspective.
Like bleeding, like blood this flying
is the weakness and the soul.
How merry, the scarlet canary, the wing on a crow.

Probably the Farmer

Probably the farmer hid in the valley
and kept his eye on his dog and son.
The English were hardly so. I think
they stood like the fools that cannot stop
rattling their basins under the sun,
opening their lungs to the dark blue fire
of the dragon boat's tongue. Probably
the farmer could believe what he heard:

I am cold. Yes I am the running drill
of ignorance that becomes sand, the disappointment
under the cliff, the eternal white claws
that intend to be kept.

Probably the farmer closed his eyes
and saw that they stood like fools
rattling their basins under morning,
opening their lungs into its fire.

Probably the farmer
had looked at the seashore and traveled
at least a mile on it, off the cliff,
not touching abalone, or mint,
or the stark or water-drunken trees;
but invented pockets to think by,
turned deliberately and walked home.

By his vision he directed his household.
If anything knew its way home he kept it;
when the seamen came they were so expected
that they killed only all the chickens,
ate only all he had.
In this way
the farmer became a citizen.

The Complex Mechanism of the Up

There is nothing ordinary in the garden.
The varieties are under the loose soil,
a swarm already stationary, don't cry
when you see it, out of gratitude.
Your flushed cheeks are looking at the sorrow
in the earth, the beautiful scent of standing.

It is hard to believe so much sadness,
don't blame yourself, you have seen it
carrying daffodils into a gong, then silence.

Listen, nothing can stop speaking,
her dress is rustling, he coughs,
couples rattle in the beams of the camera.

The sun deadens once again to no color.
If the links of the fence have an error
I am afraid I must go with them.
I must say I am grateful, and never let go.

The Cloud Parade

In deference to the cloud parade,
the horse has shed its winter red,
stamped its last horseshoe out of the shed,
and left no forwarding address.
The heavens turn furniture,
attics and beds, men with mustaches
heels over heads; they cover the sun
to a gloomy shade,
in deference to the cloud parade.

Scarves! Echoes! Pavilions!
The meat grain in bacon, the star-stun
in roast, the bone down the well, the moon
down the wane, the smoke from the fireplace,
beautifully made,
in deference to the cloud parade.

(4)

While they are young
the old provide
the moon gets full and looks down judging
not the maze of anger
but the fury
at the wasted years,
at the waste of the tender snow.
Wasted, wasted, the birds crackle,
wasted on you.

Heavy Snowfall in a Year Gone Past

Heavy snowfall in a year gone past
hammered the sudden edge
of the house foundations
to a rounder world
a whiter light after the end of day.
My favorite coat, lush sable
in color, a petty fake
that warmed me to the ears
hangs after the seasons
a beaten animal grinning buttons.
It became quite real to me
and now is matted on a hook.
How far away what mattered
has flourished without me,
along the tasty road in the wood:

clark, clark, the hidden birds call
or do wrong, do wrong, someone
do wrong, snapping apples
from out in the woodside, telling
their fathers names, pie cannonrude
barkwithfist brendanbe with cherries.

It is a vast field
where snow will fall again.
Is the vast field ownership
or a presence of mind?

As the Window Darkens

As the window darkens, as the light yearns
over the couch, as the plants drift like swans,
this is a silent poem. It will not flower
in water like a party favor, nor
will it bleed like the universe. It will
be the knot of wood that looks like a rabbit,
a knot of wood tense and growing stiffer.

The window darkens to mirror. It chimes
the reflection and what it should have been.
Have the trees and antennas seen this
blot for years, pumping like a gauge?
No wonder the birds flew away, crying out.

It is not true that the beautiful
are always false and the ugly philosophical.
But the clumsy stand as surely as the deft
at dark windows, knowing they have been deceived.
This is unbearable because we know
we see a door closing and we wait to see
that door close in our dreams, shutting us out.

How many years can come as gently as lamplight
out of the wind, each year with its own place
and the circle it made around a friend?
What the world likes is a bootstrap and locket,
work and sentiment ebbing with the light.

The ocean is a sigh at night, a dark vase
of flowers before a dark window, salt water,
someone pouring it from an abalone shell
when it first was made. It goes on forever,
a fountain, a fortuneteller in the palm of sand.
The ocean is a dish of water carried by a woman,
where the worry of our lives lies down.

House Is an Enigma

House is an enigma. It directs what is near
and what is far away. Under the north wind,
under the hill, the broad bay is blue-cold.
The cold does not matter. The whitecaps are heavy
and heavily wrapped, under the wind
in a pauper whiteness, deceptive brides
one by one drowned and pulled under.

The gull was born to a task—sorting
the park and the house from his eye, from the air:
there is no secret. Some whitecaps
reach the rocks. They thought rocks were safety.
They pound for the railroad; they reach no railroad,
no hill, no high peace on land.

But no, do not run, though you have
no instinct and small reason to be proud.
The ashes are cold and the image
is image no longer.

House alleys are rutted and glorified
by wild grasses, wild plants, California poppies.
Afternoons, mornings, behind cement-blocked
walls, the walls fluted for light and pattern
cling the vines. Garage doors are heavy wooden burdens.
The young trees, the sweet peas, the Hubbard squashes
are no burden. The rest is fence. Garbage cans.
Cats. Dogs. The great memory of the torn-away
raspberry jumble. And the cynical gravel.

I have been a poet for a time
before I hear a pot at home setting down
without finality. The spoon taps the sides
and the television glows and speaks out
in a language it says has been shattered,
and a light goes on in me; somewhere, inside.

But no, do not run, though you have
no instinct and small reason to be proud.
The ashes are cold and the image
is image no longer.

Winter Evening Poem

I remember how I came here
unraveling.
The words find a believer
and skate idly
on the frozen pond of his surface.

The ocean is given to night—
night, blue as mussel shell.
The lamp post in its quiet habit
spills light onto nothing below it.
There are no stars. We are all safe
in something better than ourselves.

The winter reached a tide
and the night was hidden
behind each winter tree
looking shy and biting its lips.
Wearing a heavy coat,
the believer holds fast.

He is night's hope,
within a hundred false springs
that are the scent of moist
brilliantine,
the sight of a limber muscle,
the light of the moon
that was last sighted
barking down the door.

A twig snaps.
Though we kneel to no stick
and fall to no knee
wouldn't you run
from what seems like a crack in the land
from a place that will not mend for you?

Happiness

The way the garden shines
through the fence slats as you pass,
the way the big moon rises
with an edge in shadow,
you see that once there was happiness.
This is the way to call it back:
Come back! Come back right away!
I am giving up neatness for you.

There is a back porch where a lonely being
faces loops of stars. Remember stars
stare back. Stars are like milk,
good for a bone, good for the teeth,
you must remember. There is a smoky room;
the tank of fish and the proletarian
smile out of the window at the Little Dipper,
getting it wrong again.

Happiness is a thread to find,
in flowers simple in the carpet,
scrolled and midnight blue.
Happiness is one lucky clover
in a hundred fields. I was afraid
when I knew who searched the fields
from the age of ten. I was afraid
when I saw the horses grazing.
But happiness
has no better argument than courage.
And my breath was decisive, passing in
and out.

Then She the Searcher parted the grass,
saw the snake whip away with a tale of her.

Amigo Acres

He looks at his egg each day.
It has grown mightier
each time he has tried to crack it.
He is told the warm air is rising.
Below the casement the damp is steaming.
The sun has come awkwardly
to its corner, through the hard clouds
that broke like dishes.

The sky has weathered the shingles.
The branches are like powerful horses
you would want to touch, balancing
a hand like a feather,
then blowing it away.
Color is a little fire.

Pale dandelion taproots are not his teeth,
but he puts them under his pillow.
His mother lives in a salad.
Night, and in the green,
great beasts of dandelion
make these hoofprints on the parking.

Shadows, cold clothes, sunlight,
glittering arrows, great goose
that festoons the dances,
yellow ducks, prosperous brothers,
trees, bright mallards diving,
all these, the bodies of the living,
rooms, rooms, houses, houses, houses . . .

Water Widow

She ripples, she weeps, follows bead bushes
down the bed of the bead stream
walking dreaming; looking to come down silk
slippery short boards of her tree house
ladder, glass as beautiful as water.
She is trilling the birds from their torn
feathers to dinner (water on a plate,
water for their armor) taking torn
clothes from speechless beasts
at the crescent tear of the stream.
Water widow, always the refrain,
she weeps, breathing in at the broken,
black, black rain.

She recalls, she remembers the day
with arms, with hands, with dangling
forests of blood madroña; she is flooding
the morning with cattle of flood,
searching the ferns, fulfilling the fears
of luggage, the fears of the violin,
the new wind of the leaf in the gutter,
the new prayer of the evangelical roots.
Black ink she weeps, with soot.
Water widow, always the refrain,
she weeps, breathing in at the broken,
black, black rain.

All of Them Who Ran from Rainclouds

The ribbons of smoke are very graceful
and continually rising. Whatever the blood
has been searching for has been found.
When I am weak, my heart flutters like a bird
in the mind that runs away from rainclouds
to lean on a glittering straw.
Someone with hair like a black squirrel
walks to a pedestal beneath a miniature carriage
and a door slams.
The birds cut the air to rags, a car chugs
and the air parts for it and parts for its shadow.
Someone wears a heart like a bandage,
like a disappointment.
Someone yawns on a lunch hour from art; and all,
all are very busy—the faces at the circus,
the whole line of buttons, the lights
passing daily over ornaments, their shadows
like nails or bird bones or like threads
weaving paper together. The refrigerator
throbs. The street is unguarded. More
petals abused all summer move imperceptibly.
They missed the snow. They never heard of it.
I have seen the snow. The snow vanishes
but wins over memory. Whatever the birds
are carving, their heads throb from it.
Whatever the plants are carving is going well.
Probably wise men for the crèche, while the cows
are carving out the manger, the children carving
angels in the snow. The whole world is busy.
All of them who ran from rainclouds.

Statue Maker

The windows are a yellow row of lights.
Indoors something is happening. Apples
are melting in the oven's sun.

"Guess what I am!"
Around and around someone swings them—
the children fall still in their game.

Where the hands end day flutters like a cloth,
throwing herself to her unity.
She says later she is the apple falling
from the apple tree, but there is no reason
to believe her. It is dusk anyway,
and soon they will go inside, nymphs
to the tree. Sirens lentamente; dear music.

The garden breathes the sounds, trusting
by the stakes in its heart. The hoe writes
in the tool shed. Rust loosens the lattice
in the scenery of sleep.

A stranger is walking
where a nymph in a statue opens
her hands to the ducks on bended knee.
In his hand, a nickle saws into the wood
of the phone booth. He is hearing
his own number back home.

Faces Passing Your Garden

When they look into your garden
the pale faces and their stances
look at your hawthorn,
your dogwood cream
skimmed from the bottle
with about the same thoughts
you expect from the gust
butting stentoriously into the young
wet leaves of the horse chestnut.
The shallow little rain is a help,
but where is the rescue? You can tell
them nothing. Nothing. For the trees
do in fact belong to you.
And only your trees
like the faces as they are.
No one sends them their bunches
of petrified trillium, no one
sends them their candygrams, no one at all.
Yet every day they must upheave
from the spell of your chrysalis,
keep leaving your hedge and your pond.

Patience Is a Leveling Thing

Patience is a leveling thing.
You want to lie down
until the weather clears.
The green wall, smoked to mesh,
has its subleties. Short words
go up hills with dust.
The patience of the line
is patience in theory.
Farms aren't without it.
Bored is the farmer,
and drawn in the jaw
while children have dynasties crashing.
He steps into different ground.
The slats are not the same;
the shade is different,
a tedious difference—like sickness,
maybe. Patience. Long words
have long lives of their own.

The Crow Is Mischief

The crow is mischief.
He is the shadow of the sun
as the owl is shadow of the moon—
wisdom and mischief, evening and dawn.
He wakes you when you sleep near the water
in a white bed.
I saw the crow, a lone shadow
high in an evergreen. This is his call
of discovery—a piece of black bread.
I pounded something with a stone.
Sometimes I think it was my heart.
Sometimes I think it was a stranger's heart,
someone now I will never know.

Writing Your Love Down

Blur and rot glare down from a heaven, from
a vest of skins where day turns it over.
Two trees have been polled: all reasons
are clever, anything startled is feigned.
Fire has surfaced again and again,
writing your love down. Oh, friendship,
see how the cold has fallen down on me!
Is it for the good
or for the worse that somewhere on a stage
someone is saying, "Beauty is made my fear"?
I would like to use the telephone.
The reason for sadness is less clear
when nothing has been said.

Baskets

The best are speckled with a color, like songs
you hum because the words are unremembered.
They say, hand me an apple though I do not
need it, then in the cold months when I need it
you will hand me an apple again. Pretend
you are sitting on the ground and a bird looks
out at you from the low part of a pine.
The color enters your basket while you look
at him. What has he said? What have you said?
What you never said is safe with another.
But how will you reach me when I am left
so far behind? I cannot weave. I cannot fly.

Bad Boats

They are like women because they sway.
They are like men because they swagger.
They are like lions because they are king here.
They walk on the sea. The drifting
logs are good: they are taking their punishment.
But the bad boats are ready to be bad,
to overturn in water, to demolish the swagger
and the sway. They are bad boats
because they cannot wind their own rope
or guide themselves neatly close to the wharf.
In their egomania they are glad
for the burden of the storm the men are shirking
when they go for their coffee and yawn.
They are bad boats and they hate their anchors.

She Will Not Dress Herself

The days come and go like waves
on pebbles. She the sea will not
wear the coat made from flame,
made from the flag of combustion.
She will not dress herself in fire.
Nor can she wear the galloping world
she dreams of. The days do not touch her,
she has the shell of a snail. She cannot
see clearly the way her wounds are shaped.
She will not dress herself, she will not dream.
She cannot see her way clearly
over the cold cinders. To sleep
she ventures out under a rowboat
to a ship at sea.
The nights come and go in the waves
beneath the ship, she will not wear
the canvas sails, she will not dress herself
in salt. The nights do not touch the sea.
The sea touches night in her rest.

Here in the Night

There is a dog barking here in the night,
here in my happiness where my neighbor
forgets all about me. The other has run
all his bath water, all the pressure that sang
in the pipes. I am waiting for my bouillon,
and while I drink it I will be happier still.
Train, go and whistle. Cars, make a sound like
a terrible wind in the huge full trees.
The leaves will spin down to the street
to be brown and to rustle, rustle, and
I am grateful for my neighbor, who forgets
all about me. The morning was worse!
When I feel bad I wonder why they do not come
to help me, why they let me go on this way,
but who are they? And what are they, but alone?
Now there is a dog barking in the night,
as if it were caring for me.

Night Typewriter Sounds

> ORMBUNKE *is a Swedish word for fern.*

They sneak through the carpet, thin stems
of invisible wire. I must get up.
The *ormbunke* hums all in silence.
Its lummox clay pot damp to sweat stands cold
in a pie plate crusted with foil,
a silver boot with a ruffle, sweet to the pen.
Stubborn but sweet. The aloe's thick bud
does not open, has no kind of stem
and no thorn. It grows formally, dark tie,
without adrenalin or terror.
The wandering Jew must be taken along,
its root furring lush in the cutting
no more than a week or so old. I must get up
without slippers to join difference.
I do not know. I am not certain.
One day I put my smart head out the door
and saw I could leave in five directions.
This is an odd house. For all we know
there is a gray thief hiding on the roof,
spreading like weather on the shingles.

Tomorrow in a Story

Needing more darkness, take the moon in hand.
Like an apple make it peel in one curl;
the constancy in the bright yellow wheat
where the moon's peel settles red like a bird
is golden among the cherries.

Crack the luna egg that does not taper.
The white joins the black pan of its moonlight.
Over the water a beautiful man lives in a boat.
The yolk will mean everything
he thought he could never reach by morning.

Tomorrow in a story she would buy
the hyacinth bulbs for the hidden pots,
the piñata, the round sample of log.
One of four. Do you differ? Do you perch?
It would be again the city, the finch;
the clerk would be still an old acquaintance.
There would be a street, another fine way
in the story of the mottled hillside.

Near her, the sun stops, looking back to sleep;
to the moon, her friend; to the pipe, her song.

She would look for baskets, a good market
for snow, for time zones, hillsides, and cities.
She would buy the bulbs and the yellow bird;
she would free the cage and step inside.

Laura Jensen

Laura Jensen was born in Tacoma, Washington, in 1948. She completed her undergraduate work at the University of Washington and received an M.F.A. from the University of Iowa in 1974.